WORKBOOK IN
HUMAN FACTORS

BARRY H. KANTOWITZ
ROBERT D. SORKIN
ROBERT J. SHIVELY
DAVID G. PAYNE

PURDUE UNIVERSITY

to accompany
HUMAN FACTORS
UNDERSTANDING
PEOPLE-SYSTEM
RELATIONSHIPS
BARRY H. KANTOWITZ
ROBERT D. SORKIN

JOHN WILEY & SONS
New York • Chichester • Brisbane • Toronto • Singapore

Copyright © 1983 by John Wiley & Sons, Inc.

All rights reserved.

Reproduction or translation of any part of
this work beyond that permitted by Section
107 or 108 of the 1976 United States Copyright
Act without the permission of the copyright
owner is unlawful. Requests for permission
or further information should be addressed to
the Permissions Department, John Wiley & Sons, Inc.

ISBN 0-471-87061-7

10 9 8 7 6 5 4 3 2 1

PREFACE

Human factors cannot be mastered solely by reading. Practical exercises are a necessary task for the serious student. Such exercises are best realized in the field by evaluating existing industrial situations, and in the laboratory by conducting careful experiments using appropriate equipment and instrumentation. It is unfortunate that few, if any, human factors courses are established with a simultaneous access to industry and to a human factors laboratory. This Workbook is not intended to replace such access, and any students who can expand their human factors knowledge in both the field and the laboratory are fortunate. Instead, this Workbook is devoted to the great majority of students who must be content with paper and pencils. Careful completion of these workbook projects will consolidate their understanding of many important aspects of human factors.

We have tried to provide a variety of projects covering both a wide range of topics and of difficulty. There is at least one project for each of the twenty chapters in the text to which this Workbook is keyed. Seven chapters have two projects for a total of twenty-seven projects. We doubt that any instructor would want to assign all twenty-seven projects in a one-semester course. Our students are most comfortable doing one project per week. But the wide range of projects allows each instructor to pick and choose those most suited to the abilities and educational goals of each class.

We have resisted the temptation to limit these projects only to those with explicit formulae and answers that are unequivocally right or wrong. As currently practiced, human factors is not black and white. There is often more than one good design and it can be difficult to determine if one good solution is significantly better than another good solution. So, while many projects have answers that are precisely correct or incorrect, many others have an intentional ambiguity. Some of the most productive and useful discussions in our classes have arisen from argument over ambiguous projects and we encourage students and instructors to entertain such healthy debate.

TABLE OF CONTENTS

Project	Chapter	Project Title	Page
1	1	Systems	1-4
2	1	Using Ergonomic Abstracts	5-6
3	2	Error Classification	7-12
4	2	Reliability	13-16
5	3	Hearing	17-20
6	3	Signal Detection Theory	21-24
7	4	Vision	25-28
8	5	Information Theory	29-30
9	5	Fitts' Law	31-34
10	6	Information Processing	35-38
11	7	Visual Display Design	39-40
12	7	Link Analysis	41-44
13	8	Auditory Displays	45-48
14	9	Speech	49-52
15	10	Controls	53-54
16	10	Tool Design	55-58
17	11	Data Entry	59-62
18	12	Feedback	63-66
19	13	Human-Computer Interaction	67-72
20	14	Decision Making and Maintainability	73-78
21	15	Anthropometry	79-80
22	16	Noise	81-84
23	17	Evaluating Floor Plans	85-88
24	17	Illumination	89-92
25	18	Airport Terminal Design	93-98
26	19	Temperature	99-100
27	20	Product Safety and Warnings	101-105

PROJECT 1

SYSTEMS

Reference Material: Chapter 1, pages 4-7 and 11-13

Objectives: The purpose of this project is to give the student some experience
    in evaluating a system in terms of the human factors considerations dis-
    cussed in Chapter 1 of the textbook.

Part I

The following passage describes some of the operations performed in
a large commercial corporation. These operations are typical of those
performed in marketing operations. Read the passage carefully and then
answer the questions in Part II.

The Alpha Supply Company (ASC) sells products that are used in industrial
applications (e.g., small hand tools, small and large power tools, safety
products, etc.). ASC's customers place orders with the company by telephoning
one of ASC's salespersons. These salespersons write the customers orders
down on order forms that are later sent to ASC's warehouse.

Occasionally a customer who is placing an order wants to know if a
specific item is currently in stock in ASC's warehouse or if the item will
have to be ordered from the factory. To check on this the salesperson uses
an intercom to talk with the warehouse employee who fills orders for that
product type. ASC's warehouse is divided into 23 "zones". Each zone stores
one type of product (e.g., zone 1 stores lubricants, zone 2 stores chemical
adhesives, etc.). Because the warehouse is organized according to product
type the salespersons must know what products are stored within each zone.
For example, if a salesperson wanted to know if there were any two lb. rubber
mallets in stock, he would have to know to call zone 16 where hand tools are
stored. The employee working in zone 16 would check on the mallets and then
pick up the intercom and tell the salesperson whether or not the items were
in stock.

When the salesperson finishes taking the customers order he looks up the
ASC product code number for each of the ordered items and writes them on the
order form. The order form is then sent out via a pneumatic tube system
(similar to the ones used in some drive-in banking operations) to the ajoining
warehouse building. The order is placed on a clipboard and the clipboard is
placed on a conveyor belt that moves continuously throughout the warehouse's

1

23 zones.

At each zone the employee working in that zone checks the order form to see if any of the items on the order form are stored in that zone. If not, the clipboard is simply placed back on the conveyor belt. If the order does call for items stored in that zone, then the employee must retrieve those items from their storage locations. All items are stored according to their special ASC product code.

In some zones the items stored there are too large and heavy to be carried by hand and must instead be retrieved using a forklift. These products are removed from their storage racks, labelled with the customers name and purchase order number, and then taken directly to the shipping room.

Employees working in zones containing smaller items use wheeled carts to collect the items specified on the orders. In addition, some specialized machines are used by the employees to fill orders. For example, a special linear counter is used in zone 4 to measure rope.

After all of the items in a zone are collected they are packaged, labelled with the customers name and purchase order number, and then placed on the conveyor belt. Finally, the employee writes on the order form which items were in stock and which items need to be ordered from the factory. The order form is then placed back on the conveyor belt.

The last stop for the products is the shipping room. ASC has 11 delivery routes, and, depending upon the customer's address, the products are placed on one of 11 racks, each corresponding to one delivery route. The products will later be placed on one of ASC's trucks for delivery the next day.

The final step in the process of filling an order is the routing of the four copies of the order form. Two copies are put on a clipboard to be given to the delivery driver and one copy is sent to the accounting office. The final copy is sent to the purchasing office so the purchasing office can order any items that are needed from the factory.

Part II

The questions that follow refer to the system for receiving and filling orders described above. Therefore the system under consideration consists only of the sales office and the warehouse. Answer all of these questions in the space provided.

A. Briefly describe the overall goals of the system described above.

_____

_____

B. What inputs and outputs are required to satisfy system goals?

_____

_____

    C. What are some of the operations required to produce system outputs? (Cite at least 10 operations.) _____

_____

_____

_____

_____

    D. What are some of the functions that should be performed by people within the system? _____

_____

_____

By machines? _____

_____

_____

    E. What are the training requirements of some of the human operators within the system? _____

_____

_____

    F. Are any of the tasks demanded by the system incompatible with human capabilities? (Yes or No) _____ Are there any tasks that seem to place an undue burden on the human operator? _____

_____

_____

3

G. What equipment interfaces does the human need to perform his job?

_____

_____

_____

     H. In what ways does the human sub-system help the machine sub-system and vice versa? _____

_____

_____

PROJECT 2
USING ERGONOMICS ABSTRACTS

Reference Material: Chapter 1, pages 21-29

Objective: This project gives the student experience in using an important
reference journal in human factors, Ergonomics Abstracts.

Part I

Each issue of Ergonomics abstracts provides an explanation of how to use
the journal and how it is organized. Completion of this project requires
that the student read and understand that section of the journal.

Ergonomics Abstracts should be located in the library at your school.
Upon locating the journal, find the April 1982 edition, volume 14, number 2.

A. Find the abstract of the article "Human Strength in the Operation
    of Tractor Pedals."

    1. What is the abstract number? _____

    2. Under what two main headings is this article listed?

        A. _____

        B. _____

    3. Under what topic does this article fall in the list of applica-
        tions?

        _____

B. Find an article on digital visual displays.

    1. What is the title? _____

        _____

        _____

5

2. In what journal did it appear? _____

_____

_____

C. What are the titles of two articles that apply to police work?

1. _____

2. _____

D. Find the abstract corresponding to number 83358.

1. What major heading is this listed under? _____

2. Why might this article be difficult for you to use? _____

_____

_____

Part II

A. List some of the problems you encountered using the classification scheme of Ergonomics Abstract.

_____

_____

_____

_____

_____

_____

_____

_____

PROJECT 3

ERROR CLASSIFICATION

Reference Material: Chapter 2, pages 32-33

Objectives: The purpose of this project is to give the student practice in classifying errors. The error classification system proposed by Swain and Guttman (1980; see textbook page 32) will be used to classify the errors described below.

Part I

Listed below are the five error categories proposed by Swain and Guttman (1980). For each of the five categories a definition of the errors that fall into that category is provided, along with an example of an error that would be classified in that category. Read each of the definitions and examples before answering the questions that follow. Note that the last three error categories (Extraneous Act, Sequential Error and Time Error) are technically Errors of Commission, but that each one is important enough and sufficiently common to justify having their own classification.

Error Category	Definition and Example
A. Error of Omission	Definition: Omitting a part of a task. Example: Not placing a filter in a coffee maker and as a result all of the coffee grounds are washed into the coffee pot.
B. Error of Commission	Definition: An action which results in performing a task incorrectly. Example: Placing your cars automatic transmission in drive when you want to go in reverse.
C. Extraneous Act	Definition: Performing a task that should not be performed because it diverts attention away from the person-machine system and thus creates the potential for damage. Example: Singing along with a song on the car radio and as a result driving past a turn you intended to make.

7

Error Category	Definition and Example
D. Sequential Error	**Definition**: Performance of a task out of the correct sequence. **Example**: Entering numbers into a calculator in the wrong order so that instead of calculating 3÷2 you calculate 2÷3.
E. Time Error	**Definition**: Performing a task too early, too late or not within the time allowed for performing the task. **Example**: Mailing your Federal Income Tax return on April 16.

Part II

Fifteen errors are described in the following section. Read each of the descriptions carefully, and then classify the error into one of the five categories defined above. Remember that an Error of Commission _may_ _possibly_ in some cases be more precisely classified as one of the last three error categories. Finally, for each error you are to briefly write down your reason(s) for classifying the error the way you did.

A. One job of warehouse employees is to remove crates from the end of a continuously moving conveyor belt. After picking up a crate the employee carries the crate a short distance, places the crate on a truck and then returns to get the next crate. Occasionally when the employee gets back to the conveyor belt he finds that a crate has fallen off the end of the conveyor belt while he was loading the last crate on the truck. These crates are usually damaged when they fall from the conveyor belt.

Error Classification: _____

Rationale: _____

_____

B. The petrochemical plant control room operator checked the hydraulic pressure gauge but forgot to record the reading on his report sheet.

Error Classification: _____

Rationale: _____

_____

C. The carpenter was sawing a piece of plywood when he heard two cars race past the house he was building. He glanced up to look at the two cars and when he looked back at his work he realized that he had cut further into the plywood than he had intended to. The plywood was ruined, so he got another piece and started over again.

Error Classification: _____

Rationale: _____

_____

D. An electrician who replaced a light switch installed the new switch upside down. In order to turn the room lights on with the new switch the switch had to be pushed down instead of up.

Error Classification: _____

Rationale: _____

_____

E. The saleswoman had just looked up a stock number in the company catalog. She was about to enter the number into her computer terminal when a fellow employee asked her what time it was. After looking at her watch and telling the person the time she went back to entering the stock number. She discovered then that she had forgotten the stock number.

Error Classification: _____

Rationale: _____

_____

F. A forklift operator wanted to back up with his load. After the forklift started to roll backwards the operator looked over his shoulder to see what was behind him. Before he could stop the forklift he had run over the tools that were on the floor behind the forklift.

Error Classification: _____

Rationale: _____

_____

G. The baggage clerk put the wrong destination tags on the passengers' luggage. The passenger arrived in Tulsa only to find that his luggage had been sent to Tuscon.

Error Classification: _____

Rationale: _____

H. The electric motor repair manual called for a washer to be put on the shaft first, followed by the gear. The repairman put the gear on first followed by the washer.

Error Classification: _____

Rationale: _____

I. A student was trying to call her professor at home, but the professor was not in. His automatic telephone answering machine told the student to leave a message at the sound of the tone. The student began her message before the tone and as a result the professor only got part of her message.

Error Classification: _____

Rationale: _____

J. While the homeowner was mowing his lawn he heard his son call to him from the neighbor's yard. While he was listening to hear what his son wanted he accidently mowed down a row of his wife's flowers.

Error Classification: _____

Rationale: _____

K. During the process of replacing the speaker on a television set the repairman forgot to connect one of the two speaker wires. This resulted in there being no sound at all from the television when his customer turned the set on at home.

Error Classification: _____

Rationale: _____

L. The bachelor loaded his automatic dishwasher, turned it on and went to work. That evening he found that his dishes were not clean. At that point he realized that he had neglected to put detergent in the machine.

Error Classification: _____

Rationale: _____

M. The legal secretary was hurring to finish photocopying a document. To save time she was pressing the START button and then placing the document as quickly as possible on the copy surface. One time the pages of the document were stuck together and the photocopier started copying before the secretary had placed the next page on the copier.

Error Classification: _____

Rationale: _____

N. In the same situation as M above, the secretary made another error. She placed the page on the copier in time, but she placed it incorrectly and only a portion of the page was copied.

Error Classification: _____

Rationale: _____

O. The lawyer wanted to tape record a political speaker so he brought a portable tape recorder to the meeting. When the lecture was over the lawyer noticed that he had inserted the microphone plug into the speaker jack when he meant to insert it into the microphone jack. As a result of this, none of the lecture had been recorded.

Error Classification: _____

Rationale: _____

_____

PROJECT 4
RELIABILITY

Reference Material: Chapter 2, pages 55-57

Objectives: The purpose of this exercise is to demonstrate to the student
the procedures for computing the reliability of serial and parallel systems.
The student is also asked to examine the effects of adding components of
less than perfect reliability to serial systems and to parallel systems.

Part I

The function of a deepsea salvage operation is to locate sunken vessels.
The system used to locate sunken vessels requires the successful functioning
of three components: (a) sonar to scan the ocean floor and display its
profile on a screen, (b) a sonar operator to detect and identify the sunken
vessels, and (c) deepsea divers to pinpoint the location of the sunken vessels.

The reliabilities of these three components are:

Component	Reliability
Sonar to scan the ocean floor	.98
Sonar operator to detect and identify vessels	.80
Divers to pinpoint the location of the vessels	.85

A. The system described above is a simple serial system. The formula for
computing serial system reliability is given in the textbook on page 55.
In the space provided below calculate the serial system reliability. (Round
all computations off to the nearest .001.)

Computations:

Reliability of the three-component serial system = _____

13

B. In the serial system examined in A. above we assumed that there were only 3 components. In the space provided, calculate the reliability of the serial system if another component (e.g., underwater metal detectors) with a reliability of .90 was added to the system.

Computations:

Reliability of the four-component serial system = _____

C. In the space provided calculate the reliability of the serial system if another component with reliability of .97 was added to the four-component serial system.

Computations:

Reliability of the five-component serial system = _____

Part II

In Part I we evaluated the reliability of simple serial systems. In Part II we will consider the situation in which several of these serial systems are linked in a parallel fashion. For example, more than one salvage ship, each with sonar, sonar operators, and divers, could be used to search for the sunken vessels.

For computational purposes each simple serial system will be considered as one component of the parallel system. Therefore, you are to use the reliabilities of the serial systems that you computed in Part 1 when answering questions in this section. Use the equation for computing the reliability of a parallel system given on page 55 in the textbook.

A. Using the reliability of the three-component serial system as the reliability of a single component in the parallel system, compute the following reliabilities:

Number of Salvage Ships Operating in Parallel	Parallel System Reliability
two	_____
three	_____
four	_____

Computations:

B. Consider the five-component serial system as a single component in the parallel system. Compute the following parallel system reliabilities:

Number of Salvage Ships Operating in Parallel	Parallel System Reliability
two	_____
three	_____
four	_____

Part III

A. What happens to serial system reliability as additional components are added to the system? _____

_____

_____

_____

B. What happens to parallel system reliability as additional components are put in parallel? _____

_____

_____

_____

C. A hybrid system is comprised of both serial and parallel arrangements of components. Under what circumstances might it be desirable to change from a strictly serial system to a hybrid system in which some components ("backups") are arranged in parallel? _____

_____

_____

_____

D. What would this switch from a serial system to a hybrid system do to the overall system reliability? _____

_____

_____

_____

16

PROJECT 5
HEARING

Reference Material: Text Chapter 3, pages 83-98

Objectives: This project illustrates some aspects of signal analysis, hearing
  thresnold, and loudness. The consequences of system nonlinearity are
  demonstrated.

I. Suppose you are given the following sound signals. Using the equation
(equation 3-5 in the text), compute the Sound Pressure Level (in dB) for each
signal. Then estimate whether each signal would be audible under quiet conditions (hearing threshold, no noise).

$$L_p = 20\log_{10}\left(\frac{p}{p_r}\right),\ p_r = .0002\mu b$$

	Frequency (Hz)	Pressure (μb)	SPL (dB)	Audible (yes/no)
A.	200	0.063		
B.	1500	0.00036		
C.	4800	0.4		
D.	8800	0.0005		

II. Three of the signals in part I would be audible in very quiet conditions. Suppose those signals were added together to form a complex sound. Plot the spectrum of the resultant sound. (Plot SPL in dB versus frequency on the graph).

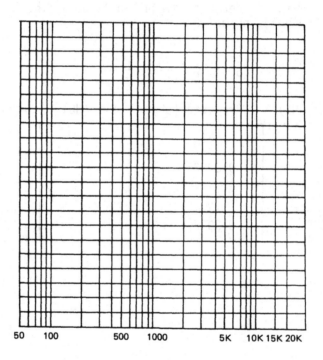

50   100   500   1000   5K   10K 15K 20K

III. Suppose this signal were passed through a linear system that amplified (or attenuated) frequencies according to the following plot. Use the plot to determine the output level of each frequency component of the signal (add the plotted value (in dB) to the input signal level).

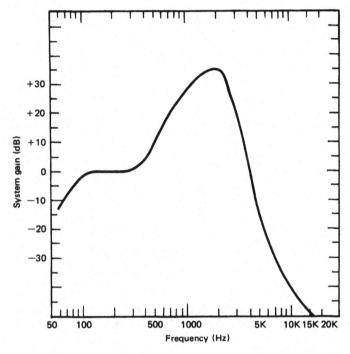

Figure 5-1 System gain function.

	Frequency	SPL input signal	System Gain	SPL output signal
A.				
B.				
C.				

D. Draw the spectrum of the output signal on the graph in II. (use a different color pen.)

IV. Suppose that the input signal in II. was passed through a system described by the following equation:

$$\text{Output} = (\text{Input})^2$$

Qualitatively describe the output signal that would now result.

V. Use Figure 3-9 in the text to estimate the loudness level, in phons, of each of the frequency components of the output signal in II.

	Frequency	SPL (dB)	Loudness level (dB phons)
A.			
B.			
C.			

PROJECT 6
SIGNAL DETECTION THEORY

Reference Material: Text Chapter 3, pages 63-83

Objectives: This project enables the student to evaluate human performance in a detection task by using Receiver Operating Curves and the d' measure. The project further illustrates how performance in a new discrimination task can be predicted from the detection data.

Two subjects were required to detect signals in a visual signal detection task. Each subject performed four separate tasks; in each task the signal was set to a different intensity (the background noise was constant). Table 1. summarizes the hit $P(Y/S)$ and false-alarm $P(Y/N)$ probabilities obtained in each condition.

Table 1. Data from the Detection Task

Signal

Subject R.D.	S_1	S_2	S_3	S_4
$P(Y/S)$	0.42	0.58	0.65	0.84
$P(Y/N)$	0.27	0.24	0.18	0.21

Signal

Subject D.E.	S_1	S_2	S_3	S_4
$P(Y/S)$	0.88	0.84	0.93	0.84
$P(Y/N)$	0.79	0.62	0.69	0.42

Part I

A. Plot the data of Table 1. on an ROC curve drawn on Fig 1.

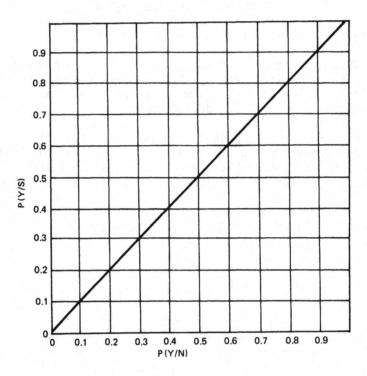

Figure 6-1 Receiver Operating Curve

B. Can you tell which subject was better at detecting the signals? Explain.

C. Which subject was more conservative about responding "yes"? Why?

II. Compute the d' value for each subject and each condition (Use Table 3-4 in the text).

	Subject	Signal Condition	calculated d'
A.	R.D.	S_1	
B.	R.D.	S_2	
C.	R.D.	S_3	
D.	D.E.	S_1	
E.	D.E.	S_2	
F.	D.E.	S_3	

Show your working diagrams and computations in this space:

III. Suppose that subjects R.D. and D.E. are going to work in a new task. In the new task there will be _two_ signals, S_2 and S_4 of the preceding experiment. On a trial of the new task, either signal S_2 or S_4 will be present in noise (the same noise as in the first experiment). The subject must report "S_2" or "S_4". This is a _discrimination_ task and performance can be predicted from the equation (3-16 in the text):

$$d'_{2,4} = \sqrt{(d'_2)^2 + (d'_4)^2 - 2\,\rho_{2,4}\,d'_2 d'_4}$$

We need to know what rho, $\rho_{2,4}$, the correlation between S_2 and S_4, is equal to. Assuming that the signals are perfectly correlated ($\rho_{2,4} = 1$) compute the predicted d' in the discrimination task. What would the d' values be if $\rho = 0.3$?

A. $\rho = 1.0$ R.D. _____

B. $\rho = 1.0$ D.E. _____

C. $\rho = 0.3$ R.D. _____

D. $\rho = 0.3$ D.E. _____

PROJECT 7
VISION

Reference Material: Text Chapter 4, pages 99-136

Objectives: To obtain experience with the concepts of visual angle and spatial
    frequency in visual acuity. To predict visual performance with an electronic
    image system. To understand the principles of color specification.

Part I
Visual Acuity

You are to arrange a work space for a visual inspection task. The task
requires the operator to make comparison judgments of surfaces having finely
detailed patterns. These patterns include short, thin lines spaced as closely
as 0.88mm.

A. What is the visual angle subtended by such a stimulus at a viewing distance
of 1.5 meters?

B. At that viewing distance, what sorts of spatial frequencies are present
in the stimulus?

C. How much of a drop in contrast sensitivity would you expect if the workers in the inspection task were drawn from a population of people who had very large (eyeglass) corrections for nearsightedness? Suggest two ways that this problem could be minimized.

Part II
Image Quality

Suppose that you are using a low-flying surveillance craft with a high resolution T.V. system. The system has 800 lines per screen and a 10° field of view. At what distance from potential targets will you be able to recognize vehicles with this system? (Use equation 4-3, Table 4-2, and assume a vehicle size of 3 meters and target contrast high enough to insure detection.)

Part III
Color

Consider the following colored light signal (the intensity at each wavelength is expressed in arbitrary units):

Colored Light Signal

Wavelength (nm)	Intensity
430	20
500	135
640	120
660	480
880	155

Use this data and Table 4-4 in the text to determine the tristimulus values needed to match each component of the light signal. For each primary, add up the values required for a match to the colored light signal. Finally, calculate the percent of each primary needed.

Tristimulus Values Needed to Match Each Component

	Wavelength	Tristimulus Values $\overline{x}\lambda$	$\overline{y}\lambda$	$\overline{z}\lambda$	Component Intensity	Primary Intensities x	y	z
A.	430				20			
B.	500				135			
C.	640				120			
D.	660				480			
E.	880				155			
						Σx	Σy	Σz

27

$\Sigma x + \Sigma y + \Sigma z = T =$

$\%x = (\Sigma x)/T \qquad =$

$\%y = (\Sigma y)/T \qquad =$

Plot the resultant primary mixture on the chromaticity diagram. Describe the hue and saturation of the light signal.

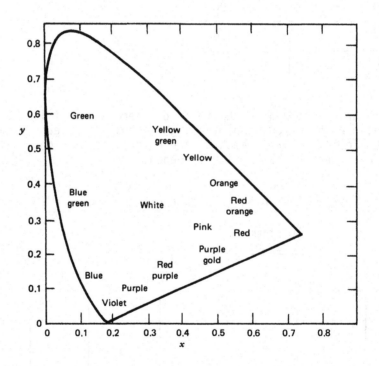

Figure 7-1 Chromaticity diagram

PROJECT 8
INFORMATION THEORY

Reference Material: Text Chapter 5 Appendix, pages 142-143, 163

Objectives: To calculate various informational statistics for equal and
unequal probability sets of events.

Part I
Univariate information theory

A. What is the amount of stimulus information in a set of 8 signal lights if
each light has an equal probability of occurrence?

B. How many bits of information would be added to A above if the number of
equiprobable lights were doubled?

C. The probabilities of the lights are changed as indicated below. Calculate
the amount of stimulus information and the redundancy.

Stimulus	1	2	3	4	5	6	7	8
Probability	.08	.25	.12	.10	.08	.05	.10	.22

Part II
Bivariate information theory

A. A five by five stimulus-response array contains the number 4 in each cell.
Calculate stimulus information, response information, joint information, and
transmitted information.

H(S) =   H(R) =   H(S,R) =

T(S:R) =

B. A five by five stimulus-response array contains the number 20 in each cell
of the main diagonal. All other cells contain 0. Calculate stimulus information, response information, joint information, and transmitted information.

STIMULUS   H(S) =   H(R) =

                                 H(S,R) =   T(S:R) =

		1	2	3	4	5
R	1	20				
E	2		20			
S						
P	3			20		
O	4				20	
N						
S	5					20
E						

C. For the blank five by five S-R matrix below insert the number 20 in only
five cells (not along the main diagonal) so that the matrix will yield the
same informational quantities you have calculated in part B above.

STIMULUS

		1	2	3	4	5
R	1					
E	2					
S						
P	3					
O	4					
N						
S	5					
E						

30

PROJECT 9
FITTS' LAW

Reference Material: Text Chapter 5, pages 154-159

Materials required: Stop watch or watch with second hand, sharp pencil, rule with mm scale, partner.

Objectives: To perform an experiment that illustrates Fitts' law. If this experiment is performed in class, pooling data will yield better results, especially if the instructor uses counterbalancing with half the class starting on target set A and half on set B.

Part I

If this is not performed in class, you will need someone to time you. For twenty seconds tap back and forth as fast as you can between the two targets in set A (or set B). Use a sharp pencil. When time is up count the dots (hits) in each target. Repeat for the remaining target set. Compute the Index of Difficulty ($\log_2 2A/W$) for each target set. What is the predicted relationship between the number of hits in set A versus set B?

ID for set A = _____ bits   ID for set B = _____ bits

(Use space below for your calculations)

Part II

Compute the mean number of hits for target sets A and B. This number is your goal. Draw two additional target sets, one with an ID 2 bits greater than target set A (target set C) and the other with an ID 2 bits less (target set D.) Again tap back and forth between targets as fast as you can until your partner says "stop." Your partner will count your hits and will say "stop" when you reach the goal. Your partner will record the time (in seconds) it took you to reach your goal. Repeat this procedure for the remaining target set and then switch roles with your partner who will tap while you time and count hits.

For each target set, including sets A and B, compute the mean movement time per hit (correct tap). Use these data to plot a Fitts' law function with ID as the abscissa and mean movement time as the ordinate.

Mean movement times

Target set A _____   Target set B _____

Target set C _____   Target set D _____

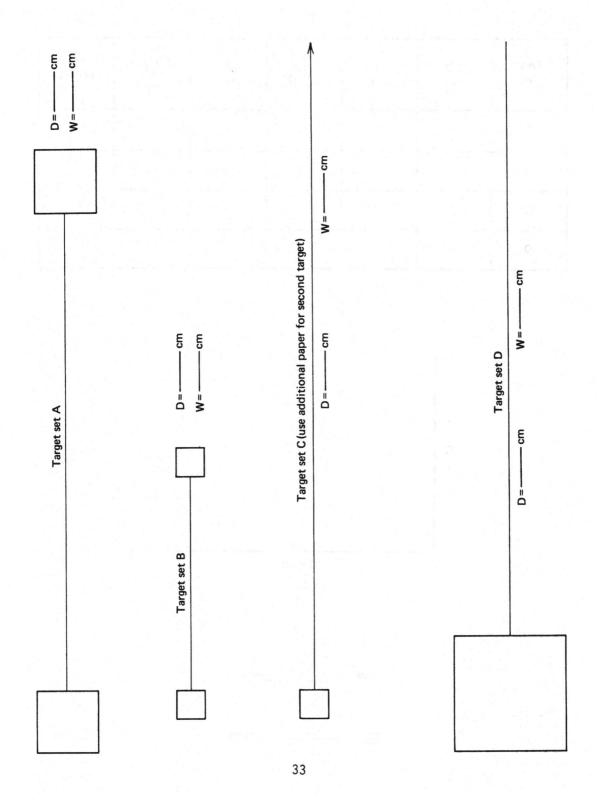

Target set A

D = ———— cm
W = ———— cm

Target set B

D = ———— cm
W = ———— cm

Target set C (use additional paper for second target)

D = ———— cm
W = ———— cm

Target set D

D = ———— cm
W = ———— cm

33

Target set	D (cm)	W (cm)	ID (bits)	MT (sec)	Number of hits	Mean MT/hit
A				20		
B				20		
C						
D						

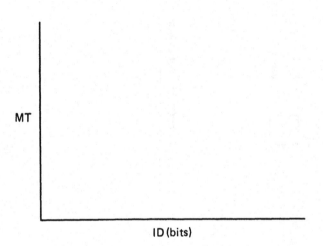

MT

ID (bits)

Slope = _____ secs/bit

$\dfrac{1}{slope}$ = _____ bits/sec

34

PROJECT 10

INFORMATION PROCESSING

Reference Material: Chapter 6, pages 168-178

Objectives: The purpose of this exercise is to demonstrate how an understanding of the basic properties of the human information processing
system can be useful in analyzing tasks performed by human operators.

Part I

A. The Postal Service has considered increasing the length of ZIP Codes
from 5 digits to 9 digits. What recommendations would you give to the Postal
Service to help make the new 9 digit ZIP Codes as easily memorized as possible?
Defend your recommendations. _____

_____

_____

B. An example of a product code number is 801301422577. Employees often
report having difficulty remembering these product codes long enough to write
them down after looking them up in a catalog.

1. What memory system is employed in performing this task? _____

_____

2. What limitations of this memory system makes this task difficult?

_____

_____

_____

3. How would you redesign this task so that it would be more compatible with human capabilities? Defend your suggestions. _____

_____

_____

_____

C. The telephone directory assistance operators in a large office building find it easy to remember new executive's telephone numbers but very difficult to remember new telephone numbers when the executives change offices.

1. Why is it more difficult for the operators to remember the changed numbers than it is for them to remember new telephone numbers? _____

_____

_____

2. What is the term used to describe this phenomenon? _____

_____

3. How could this problem be alleviated? Describe at least two possible solutions. _____

_____

_____

_____

_____

D. A telecommunications corporation is considering instituting a toll-free (800) number that could be called to obtain nationwide weather information. The human factors specialist recommended that the number for this information be 800-932-8437.

1. Is this a good recommendation? (Hint: Examine a telephone dial or keypad to determine which letters are assigned to the digits 2 through 9.)

_____

_____

_____

2. Why is this number easily remembered? _____

_____

_____

3. Are there any problems with this recommendation? Be specific.

_____

_____

Part II

A. Describe at least two types of information that you have memorized by using chunking. _____

_____

_____

_____

_____

B. Describe in some detail a mnemonic scheme that you have found useful in the past when you have needed to memorize some piece(s) of information.

_____

_____

_____

_____

_____

Name _____   Date _____

PROJECT 11
VISUAL DISPLAY DESIGN

Reference Material: Chapter 7, pages 201-222

Objectives: This project gives you practice in evaluating, modifing, and
   designing visual displays based on information from the text and lectures.

<u>Part I</u>

   Figure 1 depicts a quantitative display that the operator must read
the level of water to the exact foot. Critically evaluate the display based
upon the criterion in the text and your own thoughts on the problems of the
design. The display is a moving pointer-fixed scale with an operational
range from 0-35.

Figure 1

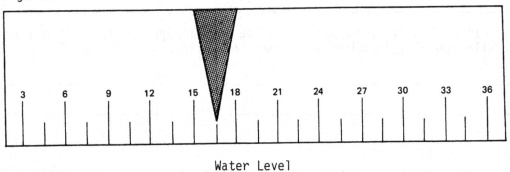

<u>Water Level</u>

A. Good Points _____

_____

_____

_____

_____

B. Bad Points _____

_____

_____

_____

_____

    If the scale in Figure 1 were to be used as a qualitative display, how might you adapt it, given the following information?
    3-12 lbs. per sqin - dangerously low
    12-24 lbs. per sqin - normal operating conditions
    24-36 lbs. per sqin - dangerously high

_____

_____

_____

_____

_____

    If the qualitative display that you just designed is to be viewed under low illumination, how might you change the design?

_____

_____

_____

_____

_____

Part II

    In the space provided design a quantitative display, in which high accuracy is needed. The range of the display will be 0 to 35 and the numbers will change very slowly.

Name   Date

PROJECT 12
LINK ANALYSIS

Reference Material: Chapter 7, pages 226-227; Chapter 18, pages 596-600

Objective: The purpose of this project is to illustrate the procedure
conducting a link analysis and give the student an opportunity to perform
a link analysis of a visual display panel.

Part I

Link analysis is a technique used by human factors designers to analyze
relationships in person-machine systems or in person-to-person systems. A
link is defined as any relationship between a person and a machine, between
one person and another, or between one machine and another. This analysis
technique provides a systematic method for optimizing the arrangement of
system links.

Link analysis has often been applied to the design of equipment layout
for an office or control room and in the layout of visual displays. However,
link analysis has also been applied to other areas. Harper & Harris (1975)
applied link analysis to police intelligence in an effort to systematically
establish relationships (links) among various individuals and organizations.
For the purposes of this project we will discuss link analysis in terms of
design of visual display panels.

Any link analysis begins with collection of information about the system
that the analysis is to be based upon. This may be accomplished through
either observation of an existing system or through simulation techniques
that give the designer an idea of how each proposed design may work in practice.

The type of link that is analyzed depends upon the situation at hand.
In this example, we will begin by looking at frequency links. These frequency
links indicate the percentage of times that an operator's eyes shift from one
display to another. Figure 1 shows the frequency links between four visual
displays. If frequency links were the only factors to be considered, the
optimal procedure would be to shorten the length of the links with the highest
values, thus decreasing the operator's eye movement time on the most used
paths.

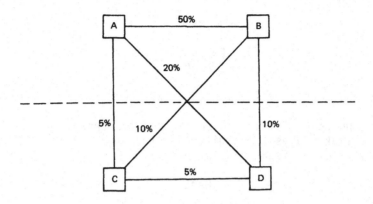

Figure 1. Frequency with which an operator's gaze shifts from one display to another.

The layout in Figure 2 minimizes the length of larger frequency link values. However, in practice other factors also need to be considered, such as the importance of each display, the time spent viewing each display and the sequence of eye shifts from one display to another. For example, although the frequency link from display A to display B represents 50% of operator eye shifts, display A may only require a scan to determine the qualitative status of the display (called a check reading). Therefore, it would be less important to locate display A directly on the line of sight, although the length of the link from A to B should still be minimized.

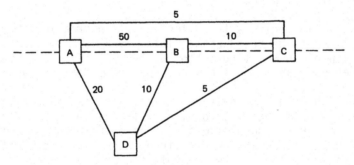

Figure 2. Frequency with which an operator's gaze shifts from one display to another

Similarly, if display C is the most important display (as rated by experts) it may demand a location on the line of sight, directly in front of the operator, even though its frequency links are low.

42

Table 1 gives additional information for each of the four displays. This information can be taken into account as shown in figure 3. Comparison of figure 2 to figure 3 will reveal how additional information allows the designer to produce a more efficient layout.

Table 1

Importance by rank	Percentage of time spent viewing each display
1 - C	A - 10
2 - B	B - 40
3 - A	C - 25
4 - D	D - 25

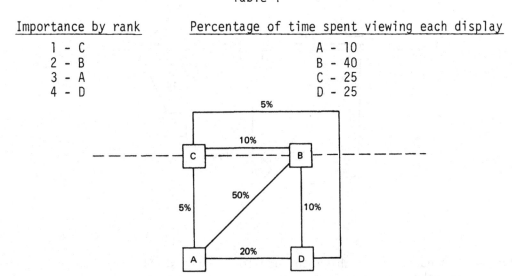

Figure 3. Frequency with which an operator's gaze shifts from one display to another.

Other designs are also possible which arrange the displays as well in terms of the information available to the designer. As more displays are added to a panel, the trade-off decisions between importance and frequency and time spent on each display become more difficult to resolve, and the number of arrangements that optimally take the information into account decrease. Part B will give you practice on the redesign of a display panel using link analysis.

Part B. A display configuration is shown in figure 4, along with the link values indicating the frequency with which the operator shifts his/her gaze from one display to another. Table 2 provides additional information about the displays. Taking this information into consideration, draw a new display configuration in the space provided. Remember, the normal line of sight is usually considered to be about 15° below the horizon.

43

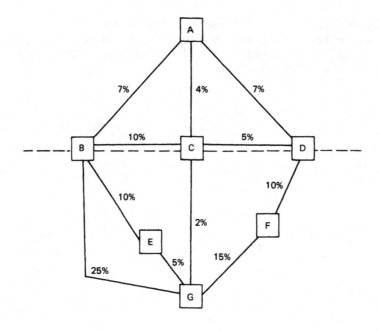

Figure 4. Frequency with which an operator's gaze shifts from one display to another.

Table 2

Display	Importance by Rank	Percentage of time spent viewing each display
A	3	10
B	5	10
C	1	25
D	2	15
E	7	15
F	6	10
G	4	15

Horizon

5°

10°

15°

PROJECT 13

AUDITORY DISPLAYS

Reference Material: Chapter 8, pages 234-240

Objectives: This exercise allows the student to apply some of the human factors principles of auditory display systems in making design recommendations for a new auditory alarm.

Part I

The emergency service vehicles (i.e., firetrucks, police vehicles and ambulances) of Metro City have auditory alarm signals that conform to the recommendations of the Committee on Hearing and Bioacoustics and Biomechanics of the National Research Council. The temporal code of these alarms is shown in Figure 1.

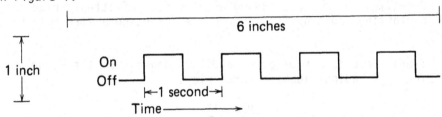

Figure 1. The temporal code of the auditory alarms used by Metro City
           emergency vehicles

While the temporal code of these various emergency vehicles is the same, their spectral characteristics vary widely. Furthermore, the sound levels of these alarms vary from 100 to 120 dB when measured at the front of the emergency vehicle where the alarms achieve their peak intensity. The alarms are less intensive in locations other than directly in front of the emergency vehicle.

At the scene of an emergency (e.g., a building fire or a brush fire) there are any number of vehicles and emergency workers. A problem at these scenes is that sometimes it is necessary to relocate a large vehicle such as an ambulance or a firetruck. When the drivers back these vehicles up it is often impossible for them to see everything in their paths. This has lead to several mishaps in which workers have been nearly run over while the vehicle was backing up.

As a result of these mishaps it has been proposed that each vehicle be equiped with some sort of auditory alarm system that would alert workers on the ground when a vehicle is operating in reverse. These alarms must be effective in a wide range of environmental settings. Furthermore it is important that the sources of these Vehicle in Reverse (VIR) alarms be easily localized by the emergency workers at the scene.

Some of the other conditions that should be taken into account when designing these VIR systems are:

A. There may be a large number of vehicles at the scene, any or all of which may have their sirens operating.

B. The emergency may be during the nighttime, or there may be a great deal of smoke, thus limiting workers' vision.

C. Most workers wear helmets at the scene of the emergency. These helmets may completely cover the head or they may have openings at the ears.

D. Sometimes workers are paying a great deal of attention to the task they are performing, such as firemen who are directing a stream of water at a fire.

E. There are a number of other auditory devices at the scene, such as communications radios, loudspeakers, etc..

Part II

In this section you are to make recommendations concerning the design of the VIR system. Answer all of the following questions.

A. In the space provided below illustrate the temporal code for your proposed VIR system, noting temporal parameters where appropriate.

B. Explain why you feel the spectral code outlined in A. above would be effective. _____

_____

_____

_____

_____

C. What recommendations would you make concerning the spectral characteristics of your proposed VIR system? On what do you base your recommendations? _____

_____

_____

_____

D. Is it feasible to incorporate a visual alarm to complement the auditory component of your VIR? If yes, describe the visual alarm. If no, explain why it is not feasible. _____

_____

_____

_____

_____

E. Besides the information provided in Part I is there any other information that would be useful to you in designing the VIR system? If so, what is the nature of this information, and why would it be useful? _____

_____

_____

_____

_____

47

F. Other than implementing an auditory VIR system on the emergency vehicles, can you think of any other ways to overcome the problem described in Part I? _____

_____

_____

_____

_____

PROJECT 14
SPEECH

Reference Material: Text Chapter 9, pages 272-305

Objectives: To compute the Articulation Index of a channel in order to predict
the intelligibility of certain speech messages on that channel; to compute
values of the Preferred Noise Criterion.

Part I

A. The level of noise in a communications channel is given by column 2 of the
Table. The peak levels of a typical speaker for this channel are given in
column 3. Compute the Articulation Index (AI) for the channel.

Center Frequency of 1/3 Octave Band	Noise Level (dB)	Peak Level (dB) of Typical Speaker	Difference	Weight	Product
200	60	71		.0004	
250	58	75		.0010	
315	55	77		.0010	
400	54	80		.0014	
500	54	82		.0014	
630	52	80		.0020	
800	50	78		.0020	
1000	56	76		.0024	
1250	56	76		.0030	
1600	65	72		.0037	
2000	65	72		.0037	
2500	68	70		.0034	
3150	70	68		.0034	
4000	65	66		.0024	
5000	70	64		.0020	

B. Describe the expected intelligibility of ordinary spoken sentences over the channel. Compare this intelligibility on the channel with that using a test of nonsense syllables.

C. Suppose the AI were better than 0.98 on two different channels. Might one of these channels be superior to the other in some tasks? Describe why (or why not) and the possible type of tasks involved.

Part II

A. What is the preferred noise criterion value (PNC) for a workplace with the following noise spectrum?

Center Frequency of Octave Bands	Noise Level (dB)
31.5	50
63	40
125	42
250	33
500	32
1000	27
2000	18
4000	16
8000	16

B. Would this PNC value be suitable for a small auditorium or a large conference room? Explain.

PROJECT 15
CONTROLS

Reference Material: Text Chapter 10, pages 321-323

Objectives: To calculate the index of accessibility for a set of controls
and to redesign the control panel to improve the index.

Part I

Calculate the index of accessibility for the control panel shown below. The
arcs show the immediate reach envelope for 25th, 50th and 75th percentile
operators. (Hint: Use s=3) The numbers inside the controls represent ranked
frequency-of-use.

The formula for the Pearson correlation coefficient is

$$r = \frac{n\Sigma xy - (\Sigma x)(\Sigma y)}{\sqrt{[n\Sigma x^2 - (\Sigma x)^2][n\Sigma y^2 - (\Sigma y)^2]}}$$

Part II

Redesign the control panel to improve the index of accessibility by relocating
controls. Cut out the controls to find better locations. Draw in your final
solution. Calculate the new improved index for your design.

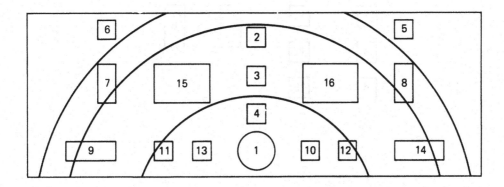

X ◄── Operator

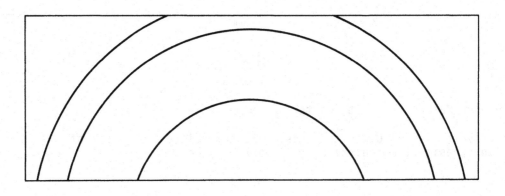

X ◄—— Operator

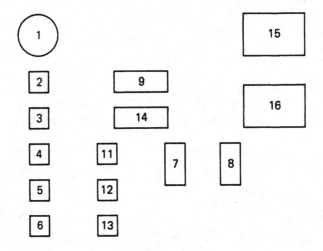

PROJECT 16

TOOL DESIGN

Reference Material: Chapter 10, pages 331-334

Objectives: The purpose of this project is to provide the student with an opportunity to apply the human factors principles of tool design in the redesign of a commonly used hand tool.

Part I

A company that builds custom interiors for recreation vehicles makes extensive use of hand tools. One of these tools is a wood saw that is used to cut out small sections of wood paneling. It has recently become apparent that there are a number of problems associated with the use of this saw. These problems are evidenced in accident reports and worker complaints about the tool.

The saw that has caused these problems is shown in Figure 1. To date there have been four major problems associated with the use of this tool.

First, several employees have cut their hands using the tool. The employees explained that their hands slipped forward off the handle and as a result their fingers were cut by the saw blade.

Second, several employees have complained of sore wrists after using the saw for extended periods.

Third, those employees who use the saw for long periods of time have also complained that after a while they find it difficult to maintain the saw blade in the proper orientation. This causes problems because if the blade is not held in the proper orientation it "catches" on the paneling, and this can cause accidents.

Finally, because the employees find it difficult to hold the blade in a constant orientation, they sometimes make inaccurate cuts in the paneling.

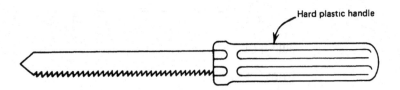

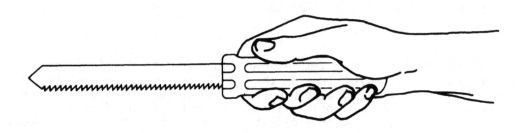

Figure 1. The handsaw described in Part I

Part II

In the space provided below, sketch a redesigned saw that you feel will alleviate the problems caused by the present design. Note any special features of your improved design.

Part III

Explain why you feel your redesigned tool is an improvement over the current design. Specifically how would your design alleviate the four problems associated with the current design?

1. Employee's hands slipping forward.

_____

_____

_____

_____

2. Sore wrists following extended usage.

_____

_____

_____

_____

3. Difficulty in maintaining proper orientation of the saw blade.

_____

_____

_____

_____

4. Inaccurate cuts caused by saw blade in improper orientation.

_____

_____

_____

_____

PROJECT 17
DATA ENTRY

Reference Material: Chapter 11, pages 336-344

Objectives: This exercise is designed to give the student experience in choosing a data entry device based upon human factors and cost considerations.

Part I

A credit card sales division of a large department store is deciding upon a computer system for issuing bills. The forty employees will spend an average of four hours per day entering data into a terminal connected to the main central computer which can then automatically issue the bills. Each employee must enter the first three categories; account no., transaction code, and amount of transaction, as shown in Table 1.

1 Account Number	2 Transaction Code	3 Amount	4 New Balance	5 Date
14732-12	1	$123.18	-$142.37	7/19/82

MAIL INVOICE? YES

Table 1. Example of a typical data entry.

This information is contained on written sales receipts. The next two categories, new balance and date, will be provided for the employee by the computer. The employee then is prompted by the computer with "Mail Invoice?", if the employee wants the invoice mailed out, as will usually be the case, the employee simply strikes the return key. The 'yes' in brackets represents a 'default' value, which the computer takes as the response if nothing else is typed in.

Visual Display Terminals

Advantages

_____

_____

_____

_____

Disadvantages

_____

_____

_____

_____

Paper Printing Terminals

Advantages

_____

_____

_____

_____

Disadvantages

_____

_____

_____

_____

Part II

A. From the information that you supplied above, recommend one type of terminal for use in this division, and explain your reasons why.

_____

_____

_____

_____

B. What considerations must be made before deciding on the type of terminal?

_____

_____

_____

C. If Visual Display Terminals are chosen, you can inexpensively display additional information. What information might be relevant?

_____

_____

_____

D. What changes could be made to the data entry process to improve it?

_____

_____

_____

_____

_____   _____
Name   Date

PROJECT 18

FEEDBACK

Reference Material: Chapter 12, pages 371-376

Objectives: The project allows the student the opportunity to evaluate and
see the advantages of a feedback system.

Part I

A. Figure 1 displays two input and output systems. The first (A) is a nonfeedback system, while the second (B) does include a feedback loop. The
equations relating input to output are:

$$A: \quad y = 10x$$

$$B: \quad y = 1000(X - \frac{y}{10})$$

where x = input, and y = output.

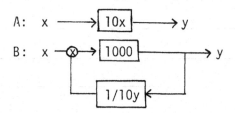

Figure 1. Input and output systems without feedback (A) and with feedback (B).

63

For the values in Table 1, determine the output produced by each system as a result of changing the forward loop gain. Assume a constant input of 100.

Table 1

Forward gain	Output(y) of A	Output(y) of B
100	1000	1000
90	900	———
80	———	———
110	———	———
70	———	———
50	———	———

B. What differences do the data reveal for feedback vs. nonfeedback systems?

_____

_____

_____

_____

_____

_____

_____

_____

_____

_____

_____

_____

_____

C. What practical significance does this have for use of a feedback system?

_____

_____

_____

_____

_____

_____

_____

_____

_____

_____

_____

Name	Date

PROJECT 19

HUMAN-COMPUTER INTERACTION

Reference Material: Chapter 13, pages 402-410

Objectives: The purpose of this project is to allow the student the opportunity to make recommendations concerning the application of the principles
of human factors in computer programming.

Part I

Maintainability of computer software refers to the extent to which the
software may be modified or altered to meet new requirments. For example,
a manager may decide that, in addition to the weekly production rate summaries
that the current software package provides, he would also like daily and
monthly data summaries. A software package exhibiting a high degree of maintainability would allow the programmer(s) to easily modify the existing software to meet these new requirements, as opposed to completely rewriting the
software to accommodate the changes.

The maintainability of a software package depends upon a number of factors.
For example, a program written and used by one person may contain ideosyncratic
terms that are meaningful to that programmer but completely nonsensical to
another programmer. Thus, one factor that determines how maintainable a
program is is who will be using and modifying the program.

Listed below are several factors that may affect program maintainability.
For each of these factors consider what types of recommendations you would
make concerning (a) pretty printing, (b) modularity, (c) comments, and
(d) mnemonic terms. You do not have to limit your recommendations to only
these four aspects, however. Finally, bear in mind that while ideally all
programs should exhibit a high degree of maintainability under a wide range
of applications, there are cost-benefit tradeoffs that accompany striving
for high maintainability.

A. Applications that require very frequent software modifications versus applications that rarely require modifications.
Recommendations: _____

_____

_____

_____

B. Software packages written and maintained by one programmer versus software written by one programmer and later modified by other programmers.
Recommendations: _____

_____

_____

_____

C. Applications in which frequent modifications are anticipated versus those in which modifications are expected to be very infrequent.
Recommendations: _____

_____

_____

_____

D. Applications in which the programmer(s) will be working with the software on a regular basis versus applications in which, once the program is written or subsequently modified, the programmer(s) will not deal with the software for quite some time.
Recommendations: _____

_____

_____

_____

E. Programs written for very specific, limited applications versus
general programs that will be modified to meet the specific needs of
different applications.
   Recommendations: _____

_____

_____

_____

   F. Very large programs (e.g., 10,000 lines) versus smaller programs
(e.g., 1,000 lines).
   Recommendations: _____

_____

_____

_____

Part II

The increasing use of computer systems has prompted human factors
specialists to study the interaction between humans and computers. The
programs of the computer system, or software, control the interaction and
thus human factors specialists have often focused on this aspect. Human
Factors in the design of software is of even more importance when the users
of the system are relatively inexperienced with respect to computer usage.
It has proven impossible to construct an algorithm to follow in the design
of software adhering to human factors principles. However, some broad
guidelines are generally agreed upon. A small sample of these guidelines
is given below. These are based upon guidelines developed by Smith (1981).
The purpose of the guidelines is to maximize the operators performance on
the system by minimizing the probability of errors. This has been called
increasing the friendliness of the system.

Human Computer Interaction Guidelines

A. Adequate documentation of the system should be provided. The documentation
   should be available on the system itself (on-line) as well as a printed
   hard copy (off-line). The on time documentation should include a 'help'
   function which contains explanations of commands, and procedures for
   correcting common errors. The user should be able to access the help
   function and then return to the same spot without loss of any programs
   or other information in the system at that time.

B. The language used to prompt the user should be clear and consistent. Esoteric computer terms should be avoided and natural language should be used instead. This is also true for error messages. For example, little information is conveyed to a novice user by the message 'error #54'. A much more informative message would be 'syntax error'.

C. The system should never allow an operator to reach a point at which the user can perform no further action. Thus, if a person is using a word processing system, then that person should not be placed in a programming mode, in which the user does not know what to do or how to return to the word processor.

D. Commands that perform critical functions, such as; save a file, delete a file, exit the program, etc., should always be confirmed before execution. This may be accomplished by the system as prompting the user for confirmation. For example, if the user types in 'Delete File A', the system should prompt the user with 'Delete File A?(Y/N)'. The system would then require a 'Y' to perform the action.

E. The user should be able to edit any entry during input. The user should also be able to check the entry, and re-enter if necessary, after input.

F. All inputs should be acknowledged by the system. After any entry the user should be informed by the system that it accepted the input.

<u>Part III</u>

A. In this section, an interactive computer system is described. In the space provided, evaluate the system using the guidelines above and other Human Factors principles. Then, provide solutions for the problems you raised.

B. A form implement supply company has an interactive computer system that was designed to process customer orders and issue requests to the factory for the needed implements. The employee using the system is given an order sheet with the implements ordered and the quantity requested. The employee is then required to enter this information into the computer system which will issue the requests to the factory. The procedure falls into three basic steps. 1) initiate system, 2) enter data, 3) terminate session. An sample interaction is given below.

   1. The operator initiates the program by logging onto the system. The user is prompted with '***', and responds by typing in their employee code number. This study the data entry program.

2. The user must enter two pieces of data for each order; model number of implement and quantity requested. After initiating the program, the user is prompted with 'A'. At this point is user inputs the model number of the implement. The user is then prompted with 'B' for the quantity requested.

3. The user terminates the session by typing in 'log'.

Sample interactions

	System prompts	User input
1.	***	2743Q
	A	7749
	B	4
	A	log
2.	***	2743Q
	A	6732
	-CODE ERROR, SESSION TERMINATED.-	
3.	***	2743Q
	A	2742
	B	3
	A	2143
	B	Ø
	-ERROR 10-	
	-SESSION TERMINATED-	

C. Evaluate the system above in terms of the guidelines provided and general human factors principles.

_____

_____

_____

_____

_____

_____

_____

_____

D. How could this system be changed to improve the efficiency?

_____

_____

_____

_____

_____

_____

_____

_____

_____

PROJECT 20

DECISION MAKING AND MAINTAINABILITY

Reference Material: Chapter 14, pages 433-436, 449-451

Objectives: To solve problem using the formulae for conditional probability, Bayes' theorem, and troubleshooting.

Part I
Conditional Probability

A. Fifty members of the pep squad went on a picnic. Ten got a tummy ache, ten got poison ivy, and five got a sun burn. Thirty returned without suffering any damage. What is the probability that a sun burned member got poison ivy? What is the probability that a member with a tummy ache did not get a sun burn? (Note - There isn't enough information to solve this problem. Specify the additional assumptions you made to solve the problem.)

B. In the dice example in the text (Table 14-1) given that $r+g > 10$ what is the probability that $r=6$?

C. A manufacturer will ship a consignment of 100 widgets if a random sample of 20 have no defects. If the consignment contains 30 defective widgets, what is the probability it will be shipped to a customer?

Part II
Bayes' Theorem

A. One urn contains 60% red chips and 40% blue chips while a second urn contains 60% blue chips and 40% red chips. A blue chip is drawn. What is the probability that the chip came from the predominantly blue urn? The chip is replaced and a second blue chip is drawn from the same urn. What is the probability that both chips came from the predominantly blue urn? (Calculate this second probability in two different ways.)

B. In a widget factory an old machine produces 30% of the widgets and a new machine produces the remaining 70%. The old machine has a defect rate of 10% versus 5% for the new machine. A widget drawn randomly from the days output is defective. What is the probability it came from the old machine? What is the probability it came from the new machine?

C. A football halfback can run through the center of the line or around the outside. The coach prefers the outside run two-thirds of the time. If she calls an outside run the halfback gains at least 5 yards 75% of the time. On an inside run the halfback gains at least 5 yards only 50% of the time. If the halfback gained less than 5 yards on the last play, what is the probability that the coach call for an outside run?

Part III
Troubleshooting

For each network and table on the next page assume all nodes are equally likely to contain the fault. Calculate the average number of tests and the standard deviation of this number for each network.

Test Sequences for Networks A

Network	If Fault Is in	Components Tested						
		1	2	3	4	5	6	
A	1	6-	5-	4-	3-	2-	1-	Probable test
	2	6+	5-	4-	3+	2-	1-	sequence and
	3		5+	4-	3+	2-	1-	test results
	4	5+		4+	3-	2-	1-	(- indicates a
	5			4+	3+	2-	1-	failed test and
	6					2+	1-	+ a satisfactory
								test)

Test Sequences for Networks B

Network	If Fault Is in	Components Tested							
		1	2	3	4	5	6	7	
B	1	1-	6-	5-	4-	3-	2-	1-	Probable test
	2		6-	5+	4-	3-	2-	1-	sequence and
	3		6+	5-	4-	3-	2-	1-	test results
	4		6+	5+	4-	3-	2-	1-	(- indicates a
	5					3-	2+	1-	failed test and
	6					3+	2-	1-	+ a satisfactory
	7					3+	2+	1-	test)

Network A

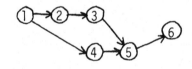

Network B

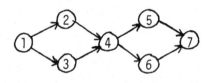

77

PROJECT 21

ANTHROPOMETRY

Reference Material: Chapter 15, pages 450-470, 494-504

Objectives: This project gives the student experience in working with anthro-
pometric data.

Part I

A.1. Use the tables in Appendix   to find the data required below on the
thump-tip reach of male USAF flying personnel.

> Thump-tip reach
> MEAN($\overline{X}$) = _____
> Standard Deviation (S) = _____
> No. of Observations (N) = _____

2. Use Table 15-2 of the text to determine the standard error of the mean.

> Standard error of the mean = _____

B.1. For the weight of male USAF flying personnel, find the following.

> MEAN($\overline{X}$) = _____
> Standard Deviation (S) = _____
> No. of Observation (N) = _____

2. Calculate the standard error of the mean for this measurement.

> Standard error of the mean = _____

C.1. Which measurement had the highest standard error of the mean? _____

2. What does the standard error of the mean tell you about a measurement?

_____

_____

_____

_____

3. Why do you think the value of one was higher than that of the other?

_____

_____

_____

_____

D. For the measurements recorded in A & B, calculate the following percentile values

	25%ile	75%ile	99%ile
Thump-tip reach -	_____	_____	_____
weight -	_____	_____	_____

Part II

A. Determine the distance away from an operator that a control could be placed so that 99 percent of the following populations could reach it. (Use Thump-tip reach).

Distance from operator

1. USAF - Females -   _____

2. USAF - Males -   _____

3. NASA astronauts -   _____

4. Italian military -   _____

5. Japanese civilians -   _____

B. What does this tell you about designing controls that will be used by a wide variety of populations?

_____

_____

_____

_____

80

_____   _____
Name   Date

PROJECT 22
NOISE

Reference Material: Text Chapter 16, pages 507-536

Objectives: To apply the principles of noise measurement for predicting the effects of noise exposure, effects on signal detection, and community reaction to noise.

Part I

A work environment contains the following sound sources:

$$L_1 = 89 \text{ dB(A)}, \quad L_2 = 83 \text{ dB(A)}$$

A. What is the total noise level?

B. Would four hours per day of that noise and four hours of 92 dBA noise exceed OSHA regulations?

81

Part II

A. You want to set the level of a brief tone (1/5 second) so that it is
easily detectable in a noisy environment. The noise spectrum level is about
42 dB in the frequency region of the tonal signal. Use text equation 16-6,
and a 10 log E/No value of 16 dB to determine the needed signal level.

B. Suppose that the tone were 2000 Hz in frequency, and 2 seconds in duration;
what should the level be (use text equations 16-7 and 16-8)?

Part III

The following data were obtained from a twenty-four hour survey of the noise in a residential area near to a small electronics plant. Samples were taken every hour.

Time	Level (dBA)	Time	Level (dBA)
0700 AM	65	1000 PM	64
0800	70	1100	66
0900	72	1200 AM	64
1000	74	0100	62
1100	70	0200	60
1200 PM	65	0300	60
100	70	0400	64
200	72	0500	68
300	73	0600	66
400	72		
500	68		
600	65		
700	64		
800	60		
900	60		

A. Compute $L_{eq(24)}$, L_d, L_n, and L_{dn} (see text equations 16-10 and 16-11).

B. What kind of community reaction would you expect to get to this noise?
Would reduction of the nighttime levels by 10 dB help much?

PROJECT 23
EVALUATING FLOOR PLANS

Reference Material: Chapter 17, pages 549-556

Purpose: The purpose of this project is to give you practice in evaluating
floor plans, based on the principles outlined in the text.

Part I

Evaluate the following floor plan based upon the human factors principles
outlined in the book. Use page 82 for your comments and be sure to discuss
good design features as well as problems with the design. Don't limit
yourself only to those design criteria discussed in the text.

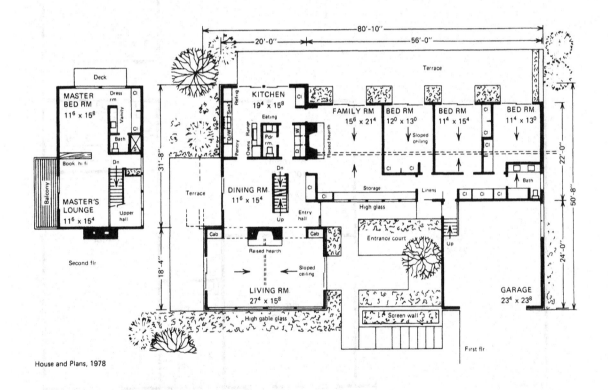

House and Plans, 1978

85

Evaluate the following floor plan, as you did in part A. Use page 83 for your comments.

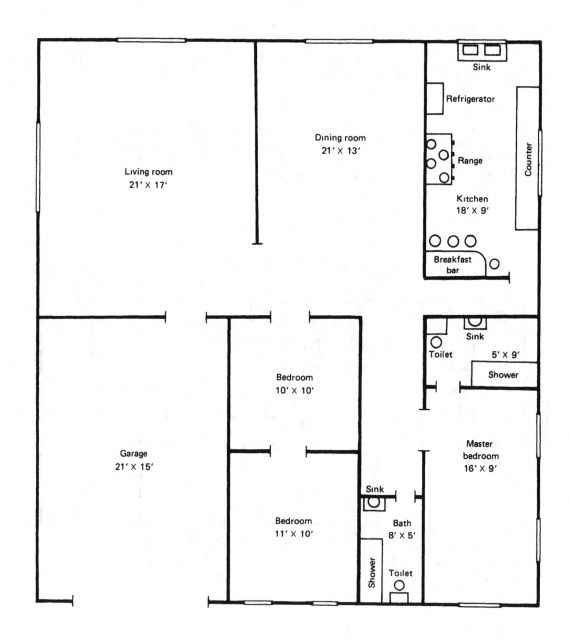

Good points _____

_____

_____

_____

_____

_____

_____

_____

_____

_____

Bad points _____

_____

_____

_____

_____

_____

_____

_____

_____

_____

_____

Good points _____

_____

_____

_____

_____

_____

_____

_____

_____

Bad points _____

_____

_____

_____

_____

_____

_____

_____

_____

_____   _____
Name   Date

PROJECT 24

ILLUMINATION

Reference Material: Chapter 17, pages 556-564

Objectives: One of the purposes of this project is to give the student
   experience in making illumination level recommendations for various
   environments. These illumination level recommendations will be made
   based on guidelines derived from the IES Procedure for Selecting Illumi-
   nances (See IES Lighting Handbook 1981 Application Volume, pages 2-20
   through 2-22.) The second purpose of this project is to give the student
   practice in calculating reflectance ratio measures.

Part I

   In this section you will read descriptions of different environments and
the activities that take place in those environments. Based on these
descriptions you will make illuminance level recommendations for these environments.

A. Large pumps are stored in a warehouse before they are shipped to customers.
The tasks performed by the warehouse employees include removing pumps from
storage racks, constructing shipping crates for the pumps, writing out packing
slips and affixing them to the shipping crates, and loading the crates on
trucks. The various pumps are all easy to discriminate from one another. The
workers, generally young to middleage men, spend most of their time doing
heavy labor (e.g., lifting pumps) and relatively little time reading orders
and writing out packing slips.

   1. Based upon this description of the task and the environment, which
one of the nine Illuminance Categories (A through I) listed in the textbook
(See Table 17-2, part I, page 560) would you assign this environment to?

_____

Why did you select this Illuminance Category? _____

_____

_____

2. What is the Illuminance Range recommended for this Illuminance Category? _____ - _____ - _____

3. Based solely upon the information from the description above, which of these three Illuminance Levels would you recommend for this environment? Briefly describe why you selected this Illuminance Level. _____

_____

_____

4. What possible additional factors might influence your recommendation?

_____

_____

B. A graphic design company is adding a new cartography (mapmaking) department. The employees who will be working in this department are all recent college graduates. Due to the nature of the work these employees must often work with very small visual displays, although not all of the work involves minute details.

1. Which Illuminance Category (See Table 17-2, part I) would you assign the cartography department to? _____
Why did you select this Category? _____

_____

2. What are the low, medium, and high Illuminance Levels recommended for this Illuminance Category? _____

_____

3. In order to select which of these three illuminance values to recommend you need to consider the workers age, the task requirements for speed and/or accuracy, and the reflectance of the task background. Task background refers to that portion of the task upon which the meaningful visual display is presented. (e.g., In the project you are now working on the paper is the task background and the letters and words are the visual display.) You will use a weighting system to take these three factors into account when making your Illuminance Level recommendation.

A. Assign a weighting factor of -1 if the workers are under 40 years, a 0 if they are 40-55 and a +1 if they are over 55.

Age weighting factor = _____

B. If speed and/or accuracy are not important assign a weighting factor of
-1. If speed and/or accuracy are important assign a 0 and assign a +1 if
they are critical in this task.

Speed/accuracy weighting factor = _____

C. If the task background has very high reflectance (i.e., greater than 70%)
assign a -1 weighting factor. A 0 weighting factor should be assigned if
the task background reflectance is moderate (30%-70%) and a +1 weighting
factor if the task background reflectance is very low (less than 30%)

Task background reflectance weighting factor = _____

D. What is the algebraic sum of these three weighting factors? _____

E. The sum of the three weighting factors will now be used to determine which
Illuminance Level to recommend.

If the sum of the weighting factors equals	The Illuminance Level to Recommend is
-2 or -3	Low Level
-1, 0, or +1	Medium Level
+2 or +3	High Level

What is the Illuminance Level (in footcandles) that you recommend?

_____

F. What Illuminance Level would you recommend if the same employees were still
on the job 30 years from now? _____

G. What Illuminance Category would you assign to a circulation desk at a
library? Use your knowledge of what activities take place at a circulation
desk to make your recommendation. _____

Why did you select this Illuminance Category? _____

_____

_____

91

What Illuminance Level would you recommend for the circulation desk at a public library where the workers are mainly senior citizen volunteers?

_____ footcandles

Part II

A. Listed below are the illumination values for three different light sources. Use these values to calculate the luminance that would be obtained for the two reflectance values given.

Light Source	Illumination Level	Luminance with Reflectance = .45
A	50 footcandles	_____ footlamberts
B	750 footcandles	_____ footlamberts
C	15 footcandles	_____ footlamberts

Light Source	Illumination Level	Luminance with Reflectance = .85
A	50 footcandles	_____ footcandles
B	750 footcandles	_____ footcandles
C	15 footcandles	_____ footcandles

B. Assuming a reflectance value of .80 what amount of illumination would be needed to obtain the following luminance levels? Assume the surface area to be 1 square foot.

(1) 1000 footlamberts requires _____ footcandles of illumination

(2) 600 footlamberts requires _____ footcandles of illumination

(3) 360 footlamberts requires _____ footcandles of illumination

(4) 1200 footlamberts requires _____ footcandles of illumination

Name	Date

PROJECT 25

AIRPORT TERMINAL DESIGN

Reference Material: Chapter 18, pages 570-578

Objectives: The purpose of this exercise is to give the student experience
in evaluating the advantages and disadvantages of centralized and decentralized airport passenger terminals.

Part I

One of the most fundamental decisions that must be made when designing
an airport passenger terminal is whether to construct a centralized or a
decentralized terminal. In this section you are to evaluate the relative
advantages and disadvantages of centralized versus decentralized terminals
in terms of the facilities that are required at the passenger terminal. You
should consider both construction and operating costs as well as convenience
to the passengers and visitors using the terminal. Listed below are five
major types of facilities required at a passenger terminal (Ashford and
Wright, 1979, pages 235-237), along with a brief description of the required
facilities. Below each description list the advantages and disadvantages of
centralized designs and of decentralized designs.

A. Access and the landside interface. Facilities must be provided to
accommodate a transfer of passenger flow from the available access modes to,
from, and through the terminal. These facilities include: curbside loading
and unloading, curbside baggage check-in, loading and unloading areas for
buses, taxis and various rapid surface modes of transportation.

Centralized	Decentralized
Advantages: _____	Advantages: _____
_____	_____
_____	_____
Disadvantages: _____	Disadvantages: _____
_____	_____
_____	_____

93

B. <u>Processing</u>. These facilities include airline ticketing and passenger check-in, seat selection, baggage check-in, baggage claim, gate check-in (where desirable), and security check areas (where desirable).

<table>
<tr><td><u>Centralized</u></td><td><u>Decentralized</u></td></tr>
<tr><td>Advantages: _____</td><td>Advantages: _____</td></tr>
<tr><td>_____</td><td>_____</td></tr>
<tr><td>_____</td><td>_____</td></tr>
<tr><td>Disadvantages: _____</td><td>Disadvantages: _____</td></tr>
<tr><td>_____</td><td>_____</td></tr>
<tr><td>_____</td><td>_____</td></tr>
</table>

C. <u>Holding Areas</u>. These facilities must accommodate the passengers and visitors while they are not in the processing areas. Furthermore, it is in these holding areas that a considerable portion of airport revenues are generated. The facilities that may be required include: passenger lounges, passenger servive areas (e.g., washrooms, public telephone, information, first aid, barber shop, beauty parlor, etc.), concessions (e.g., bars, restaurants, banks, car rentals, etc.), and observation decks and visitor lobbies.

<table>
<tr><td><u>Centralized</u></td><td><u>Decentralized</u></td></tr>
<tr><td>Advantages: _____</td><td>Advantages: _____</td></tr>
<tr><td>_____</td><td>_____</td></tr>
<tr><td>_____</td><td>_____</td></tr>
<tr><td>Disadvantages: _____</td><td>Disadvantages: _____</td></tr>
<tr><td>_____</td><td>_____</td></tr>
<tr><td>_____</td><td>_____</td></tr>
</table>

D. <u>Internal Circulation and Airside Interface</u>. Internal circulation includes corridors, walkways, people movers, ramps and tramways. Airside interface includes loading facilities (e.g., jetways, stairs, etc.) and mobile lounges.

<u>Centralized</u>	<u>Decentralized</u>
Advantages: _____	Advantages: _____
_____	_____
_____	_____

E. <u>Airline and Support Activities</u>. Some of the required facilities include airline offices, passenger and baggage processing stations, storage for wheelchairs and carts, airport management offices, offices for security personnel, public address systems, flight information, and maintenance personnel offices.

<u>Centralized</u>	<u>Decentralized</u>
Advantages: _____	Advantages: _____
_____	_____
_____	_____
Disadvantages: _____	Disadvantages: _____
_____	_____
_____	_____

Part II

Shown on the next page are four types of passenger terminal designs, one decentralized design (Linear or Gate Arrival) and three centralized designs (Transporter, Central Terminal with Remote Satellites, and Central Terminal with Pier Fingers). In this section you are to evaluate these designs with respect to differing passenger needs.

95

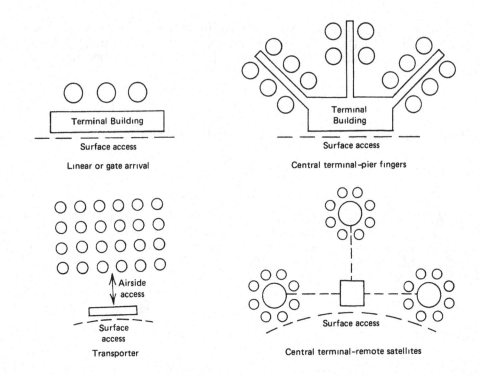

Figure 1. Four examples of passenger terminal designs. (Adapted from Ashford, N. & Wright, P. H., Airport Engineering, New York: John Wiley & Sons, 1979.)

A. Which design would be best for an airport in which most passengers use the airport as either an origin or a destination, and relatively few passengers transfer from one plane to another? Support your answer.

_____

_____

_____

B. Which design would be least desirable for an airport in which there were large daily and/or seasonal variation in passenger loads? Why?

_____

_____

_____

C. The figure below shows the relationship between the percent utilization and the relative costs for a hypothetical airport for both a transporter system and a fixed gate system (e.g., a central terminal with pier fingers).

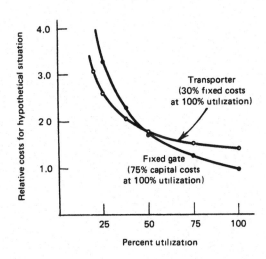

Figure 2. The hypothetical relative costs of fixed and mobile gates as a function of percent utilization. (Adapted from DeNeuville, R. Airport systems planning: A critical look at the methods and experience, Cambridge, Mass: The MIT Press, 1976.)

Assuming the projected utilization for airline A was 25%, which terminal design would be more economical for airline A? _____

_____

_____

If airline B projected a 75% utilization rate, which terminal design would be best for airline B? _____

_____

_____

D. Assume the projected utilization rates for airlines A and B given in problem C above. Design a terminal that would accomodate the needs of both airlines. Illustrate your design in the space below. Be sure to label the illustration.

E. Explain why your design meets the needs of both Airline A and Airline B. _____

_____

_____

_____

_____

_____   _____
Name   Date

PROJECT 26
TEMPERATURE

Reference Material: Chapter 19, pages 607-617

Objectives: This exercise allows the student to work with various indexes
   of performance decrement in response to heat.

Part I

Figure 19-6 in the text displays three heat tolerance curves. From this
figure, determine the maximum temperature at which a worker could work for
two hours without decrements in mental performance for the curves labeled
Hancock and Wing.

Max temp at 2 hours without decrement

A. Hancock = _____

B. Wing = _____

C. Why do these estimates differ? _____

_____

_____

_____

_____

Part II

Use the arithmatic mean of the Hancock & Wing estimates in part II.
Iso-decrement curves for mental reaction time and for vigilance and complex
tasks are displayed in Figure 19-7. For a two hour duration which Iso-
decrement curve does the mean of the estimates fall closest to:

A. For Mental Reaction Time _____

B. For Vigilance and complex tasks _____

    C. At what temperature would the decrement in vigilance and complex tasks be equal to the decrement experienced at the present temperature for mental reaction time? _____

    D. What is inadequate about these types of performance decrement estimates? _____

_____

_____

_____

__ _____

_____

_____

    E. How might these estimates be improved? _____

_____

_____

_____

_____

_____

_____

PROJECT 27
PRODUCT SAFETY & WARNINGS

Reference Material: Chapter 20, pages 624-635

Objectives: This exercise gives the student practice in evaluating situations
  that could possibly result in litigation.

Part I

     The plaintiff in a product liability case can file under several legal
principles: negligence, strict liability, implied warranty, express warranty
& misrepresentation. Each of these principles is defined in the text along
with examples. For each of the following scenarios decide 1) Who is the
defendent?, 2) Which principle, previously mentioned, is involved?, 3) What
could the company/individual have done to avoid liability?

A. Airline A uses a traditional navigation system that has been successful
in the past. A new more sophisticated and accurate system is available at
a higher cost and most other airlines have switched to the new system. During
a storm, a flight of airline A hits a mountain. The follow-up investigation
finds that the navigation system was defective and the cause of the accident.

1. Who is the defendent? _____

_____

_____

_____

2. Which principle, previously mentioned, is involved? _____

_____

_____

_____

3. What could the company/individual have done to avoid liability? _____

_____

_____

_____

B. A person purchases a vegetable slicer for use in the home. The slicer carries the warning "Caution! Blades are Sharp. Keep fingers away from blades & keep this instrument (utensil) away from small children." When slicing cucumbers for a salad, this person slices down to the end of the cucumber and on the next pass over the blades slices his/her hand.

1. Who is the defendent? _____

_____

_____

_____

2. Which principle, previously mentioned, is involved? _____

_____

_____

_____

3. What could the company/individual have done to avoid liability? _____

_____

_____

_____

C. A typist has removed the top cover from her typewriter. There is a warning label inside that reads "Warning: Do not operate with cover off." While keeping her eyes on the page she was typing, her hair was caught by the carriage as it moved along. Her hair had to be cut out to free her.

D. The same situation described in 'C' occurs. However, when the cover is removed the carriage is on the far right. In this position, it covers the warning label.

1. Who is the defendent? _____

_____

_____

_____

2. Which principle, previously mentioned, is involved? _____

_____

_____

_____

3. What could the company/individual have done to avoid liability? _____

_____

_____

_____

E. A couple is travelling along an interstate highway at 60 m.p.h., 5 m.p.h. above the legal speed limit. As the wife, sitting on the passenger side, shifts posture, she inadvertantly strikes the floor mounted automatic transmission control with her knee. This pushes the transmission into reverse causing the car to immediately lock its tires causing a skid which results in the car rolling over into its top.

1. Who is the defendent? _____

_____

_____

_____

2. Which principle, previously mentioned, is involved? _____

_____

_____

_____

3. What could the company/individual have done to avoid liability? _____

_____

_____

_____

F. A car is under written warranty for 1 year for major mechanical failure.
The owner of the car has performed all necessary maintenance to keep the
warranty valid. After owning the car for 3 months, the power steering went
out when the owner was turning a corner causing the car to ram into a house.

1. Who is the defendent? _____

_____

_____

_____

2. Which principle, previously mentioned, is involved? _____

_____

_____

_____

3. What could the company/individual have done to avoid liability? _____

_____

_____

_____

104

G. The same situation in F occurred after the owner had the car for 2 years.

1. Who is the defendent? _____

_____

_____

_____

2. Which principle, previously mentioned, is involved? _____

_____

_____

_____

3. What could the company/individual have done to avoid liability? _____

_____

_____

_____